100 BLACKBOARD PROBLEM-SOLVING ACTIVITIES

by Les Landin and Linda Hagelin

Fearon Teacher Aids

We dedicate this book to all students
who enjoy being challenged.

This Fearon Teacher Aids product was formerly manufactured and distributed by American Teaching Aids, Inc., a subsidiary of Silver Burdett Ginn, and is now manufactured and distributed by Frank Schaffer Publications, Inc. FEARON, FEARON TEACHER AIDS and the FEARON balloon logo are marks used under license from Simon & Schuster, Inc.

Editorial Director: Virginia L. Murphy
Editor: Kristin Eclov
Illustration: Teena Remer
Cover and Inside Design: Rose Sheifer

Entire contents copyright © 1994 by Fearon Teacher Aids, 23740 Hawthorne Blvd. Torrance, CA 90505. However, the individual purchaser may reproduce designated materials in this book for classroom and individual use, but the purchase of this book does not entitle reproduction of any part for an entire school, district, or system. Such use is strictly prohibited.

ISBN: 0-86653-922-0

Printed in the United States of America
1.19 18 17 16 15 14 13 12 11

Introduction

Introduce basic critical-thinking and problem-solving skills to your classroom using these simple blackboard activities. There are one hundred blackboard activities provided to help your students practice problem-solving skills, such as sequencing, prioritizing, brainstorming, predicting outcomes, and so on.

The blackboard illustrations can be copied or easily drawn. They are not meant to be cartoon creations as these activities are not art lessons. If your students laugh at your drawings and say "What is that?" Tell them it's a boat or a giraffe. Not only will you have your students' attention, but you have added some humor to teaching problem-solving.

Have fun adapting these problem-solving activities to meet the interest and ability levels of your students.

GLOSSARY

There are different approaches for teaching critical thinking and problem-solving. We feel it is important to have a common understanding of terms. Therefore, the terms below give a general explanation of the concepts used in this book.

Adapting: A flexible reinterpretation of a problem and solution after the problem-solving has begun.

Application: Taking action to apply a solution to a problem.

Creativity: Building original ideas that lead to unique solutions. This may involve individual invention or team brainstorming.

Decision-Making: Evaluating, prioritizing, and predicting the outcome.

Elaborating: Embellishing or improving solutions to a problem.

Modifying: Changing and adapting ideas to improve solutions to a problem.

Perception: Awareness or recognition of a problem.

Problem: A puzzling situation presented in a verbal or graphic context.

Recall: Memorizing or remembering knowledge related to a problem.

Solution: Solving a problem within a verbal, graphic, symbolic, or physical framework.

Understanding: Comprehending the parts and general nature of a problem.

Contents

Friends 6
Hopscotch 8
Ladder 10
The Dragophant 12
Barnyard 14
Rover 16
Shorty Giraffe 18
Around the World 20
Tree Fort 22
Commercial Conversion 24
Airmobile 26
Shipwrecked 28
The Kid Next Door 30
The Queen's Guard 32
Across the Tracks 34
Out the Door 36
Cloud Nine 38
The Chase 40
Musical Search 42
Obstacle Course 45

FRIENDS

A Primary Activity to Introduce
Team Brainstorming and Evaluation

1. FRIENDS: Draw the above picture on the blackboard.

2. Explain the following situation to the children: Hisakazu's family recently moved to Chicago from Japan. Today is the first day of school, Hisakazu is nervous about attending a new school in a new country. He doesn't have any friends. Before school each day, the children play on the playground. Hisakazu is unhappy because he does not know how to play the games the other children are playing.

    Ask the children how they would feel if they were going to school in a different country. Encourage the children to discuss all the changes that take place in moving to another country, such as language, culture, making new friends, and so on. Write the children's responses on the blackboard.

3. Divide the class into two teams. Team One should think of ways Hisakazu could become more familiar with the customs of the United States. How could Hisakazu make new friends? Have Team Two think of ways the other children could make Hisakazu feel more welcome.

4. Discuss each team's ideas for solving Hisakazu's problems. Record the responses on the blackboard. Have the class vote on the best solutions.

5. FRIENDLY GAMES: Divide the class into teams of four or five children. Ask the children how they would solve the following problem: Hisakazu is standing by himself watching several children play "Statue Tag" on the playground. Hisakazu doesn't know how to play the game.

Encourage each team to take turns acting out how they would solve Hisakazu's problem. Record the solutions on the blackboard.

6. MAKING FRIENDS: Encourage the children to talk about their own experiences with being new in school or being the only person who doesn't know how to do something. Discuss how hard it can be to make new friends. Remind the children of how important it is to be respectful of other people's feelings. Have the children think of ways to be respectful to each other in the classroom, such as saying "please" or "thank you," not interrupting when someone is talking, and so on. Record the children's responses on the blackboard. Encourage the class to practice being respectful at school and at home.

9. FRIENDLY MOODS: Draw the faces below on the blackboard. Give each child a piece of paper. Ask the children to copy the faces on their papers. Have the class label the various moods each child is experiencing, such as anger, happiness, and so on. Invite each child to draw a face that shows how he or she is feeling that day. Ask volunteers to come to the front of the room, have the class guess the mood of each volunteer from his or her face drawing.

HOPSCOTCH

**A Primary Activity Encouraging
Physical Movement and Pattern Recognition**

1. HOPSCOTCH: Draw the above hopscotch pattern on your blackboard.
2. Provide children with 9 paper squares 2" x 2" (5.0 cm x 5.0 cm) in size.
3. Explain that you will be playing hopscotch, but only your fingers will do the hopping.
4. Ask the children to arrange the paper squares in hopscotch patterns on their desks.
5. Once the squares are in place, demonstrate how to use fingers to hop around the squares. Have the children create their own hopping patterns. Ask several children to share their patterns on the blackboard.
6. HOPSCOTCH SHUFFLE: Ask children to arrange their squares into different patterns. Have several volunteers draw their new patterns on the blackboard. Encourage the children to make as many patterns as they can.

7. **HOPSCOTCH BY THE NUMBERS:** Number the hopscotch squares from 1 to 9. Rearrange the squares in different patterns on the blackboard. Encourage the children to repeat the new patterns using the paper squares.

8. **PICTURE HOPSCOTCH:** Divide the class into teams of four or five children. Give each team member 7 paper squares. Encourage the children to work as a team to create a picture using all of their squares. Give the teams 30 minutes to work. Have each team draw their picture on the blackboard.

9. **GEO HOPSCOTCH:** Draw several examples of geometric shapes on the blackboard. Show the children how to create various geometrical figures by folding or overlapping the squares. For example, demonstrate how to make a triangle by folding the square in half, matching corner to corner. Encourage the children to experiment to create their own geometric shapes.

LADDER

A Primary Activity Introducing
Team Brainstorming, Idea Creation, and Prioritizing

1. LADDER: Draw the above scene on the blackboard.
2. Explain the following situation to the children: Robert accidentally threw his ball on the roof. He found a ladder in the garage, but the ladder was too short to reach the roof.
3. Ask the children to think of as many solutions to Robert's problem as possible.
4. Divide the class into four teams. Have each team discuss their solutions to Robert's problem.
5. Encourage each team to choose a spokesperson to present their best solution to the class. Write the responses on the blackboard. Ask the teams to explain how they came up with their answers.
6. UP THE LADDER: Give the class the following problem: The roof is 10 feet (3.1 m) high and the ladder is 6 feet (1.8 m) long. How long should the ladder be to reach the roof? Write the children's answers on the blackboard. Feel free to change the dimensions to make the problem more challenging.

7. TOOLS AND A LADDER: Ask the children to think of all the tools and materials they need to get the ball off the roof. List the tools on the blackboard. Have the children number the tools in order of importance.

8. STEP LADDER: Have the children discuss the steps Robert should follow to get the ball off the roof. List the steps on the blackboard. What is the first thing Robert should do to get the ball off the roof?

9. ADVANCED LADDER: Draw a ladder on the blackboard. Do not draw all the rungs on the ladder. Explain that the rungs on the ladder are 18 inches (45.7 cm) apart. How many rungs will the ladder need to reach a 9 foot (2.7 m) roof? (You will need 6 rungs to reach the roof.)

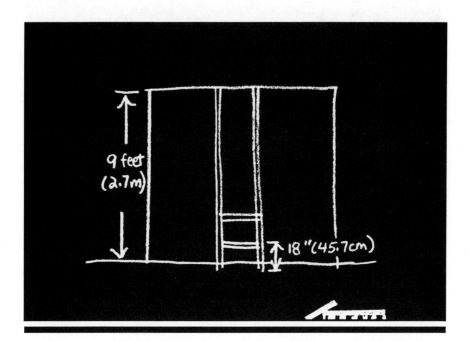

THE DRAGOPHANT

A Primary Activity Encouraging Predicting Outcomes and Elaborating on Humorous Ideas

1. THE DRAGOPHANT: Draw the above picture on the blackboard.
2. Explain that an elephant married a dragon. Their baby was part elephant and part dragon. This animal was called a dragophant. The dragophant was a very happy baby, but as he grew older, his parents discovered a problem. Whenever the young elephants bathed in the river, they would blow water at each other with their trunks. The dragophant tried to be like the other elephants and blow water with his trunk, but dragon fire came out instead. Soon all of dragophant's friends were afraid to play with him. What could he do?
3. Ask the children how to solve the dragophant's problem.
4. List the children's ideas on the blackboard. Write each child's name next to his or her response.
5. Give praise or points for the most unusual and humorous ideas. Announce the winners and discuss why their ideas were especially unique.

6. DRAGOPHANT FRIENDS: Read examples of mythical animals in literature to the children. For example, Rudyard Kipling's "Just So Stories" or "Aesop's Fables" would provide a wonderful background for further discussion of unusual animals, such as the dragophant. Record different mythical animals on the blackboard. Discuss what it would be like to be a mythical animal. Encourage the children to think of interesting characteristics of mythical animals, such as a fire-breathing elephant.

7. DRAGOPHANT'S COUSINS: Ask the children to think of unusual animal combinations, such as the dragophant or a jackalope. Have each child draw his or her creation on the blackboard and explain the animal combinations.

8. DRAGOPHANT DRAWLS: Have the children create sounds their mythical creatures would make. Encourage each child to name his or her animal. Write the animals' names on the blackboard. Record each child's name next to his or her animal. Choose one child to call out an animal name from the list. The child who represents that animal should make its sound.

9. HELPFUL DRAGOPHANT: Ask the children to write a list of the useful purposes for dragophant's unusual fire-breathing trunk. Record the children's ideas on the blackboard.

BARNYARD

A Primary Activity to Help Children
Generalize and Recognize Relationships

1. BARNYARD: Write the following farm animal names—cat, dog, cow, and pig—on small pieces of paper. Pass out the papers so each child receives one of the four farm animals. Remind the children not to share their animals with their neighbors.

2. Write these animal names on the blackboard—cat, dog, pig, and cow. Discuss with the children the different sounds each animal makes. Write the sounds next to the animals on the blackboard. Explain to the children that only when each child's animal name is pointed to, should that child make the animal sound. For example, when the children see the word *cow*, each child with a cow piece of paper should say "Moo."

3. After you have written the animal names on the blackboard, encourage each child to find the other farm animals in his or her group.

4. Encourage each group to make their animal sound, in unison, when their group's name is mentioned. Have the children in each group go up to the blackboard at the same time and write their animal's name.

5. BEASTY BARNYARD: Encourage each group to think of different characteristics for their animal, such as what a cow eats or does a cow run fast. Write each group's ideas on the blackboard.

6. BARNYARD NOISES: Ask the children to read the list of animals on the blackboard. Discuss the sound each animal makes. Record the animal sounds in random order on the blackboard. Have the children match the animal to the sound it makes.

7. BARNYARD BUDDIES: Encourage the children to name their favorite animals. Write the list on the blackboard. Invite the children to draw their three favorite animals. Display the drawings on the blackboard.

9. BARNYARD SHUFFLE: Write the following list of animals on the blackboard—zebra, horse, elephant, leopard, rhinoceros, cow, rooster, tiger, monkey, cat, bear, wolf, turtle, dog, moose, lion, chicken, and giraffe. Encourage the children to identify all the animals that live on a farm. Ask a volunteer to come up to the blackboard and circle each barnyard animal. There are a total of six farm animals. Discuss where the other animals live. Point out on a map where several animals on the list live, such as tigers live in India.

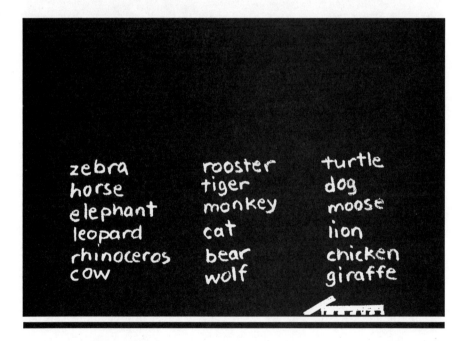

ROVER

**A Primary Activity for
Decision-Making and Predicting Outcomes**

1. ROVER: Draw the above picture on the blackboard.
2. Explain the following situation to the class: Rover, the dog, is in a boat that is leaking. To make matters worse, Rover is afraid of the water and cannot swim.
3. Ask the children to think of tools that would help Rover keep the boat from sinking. Encourage the children to be creative.
4. Ask each child to name the tool he or she chose and explain why. Record the children's responses on the blackboard.
5. ROVER'S PADDLE: Ask the children to draw pictures of their tools on the blackboard. Encourage the class to guess how each tool can be used.
6. ROVER'S ADVENTURES: As a class, write a story about Rover's adventures. Encourage the children to predict what's going to happen next to Rover. Write the story on the blackboard. Encourage the children to copy the story on hand-writing paper. Invite each child to draw an illustration for his or her story page. Combine the stories into a class booklet.

7. **ROVER UP THE CREEK:** Discuss why a leaking boat sinks. Divide the class into groups of four or five children. Give each group a small container of water and a piece of clay. Challenge the children to build boats that float. Have each group draw a picture of their boat on the blackboard. Time how long each clay boat stays afloat. Record the times on the blackboard. After the boats have been tested, compare the pictures on the blackboard to the actual boats.

8. **ROVER'S CREW:** Ask the children what would happen if another animal was in the leaky boat instead of Rover, such as a lion or a beaver. Draw the picture below on the blackboard. Encourage each child to choose an animal to take Rover's place in the leaky boat. Invite the children to write stories about their animals' adventures. Ask volunteers to share their stories with the class.

SHORTY GIRAFFE

**A Primary- or Intermediate-Grade Activity to
Encourage Brainstorming, Original Ideas, and Humor**

1. SHORTY GIRAFFE: Draw the above picture on the blackboard.
2. Tell the class the following story: Shorty was a very short giraffe. All of the other giraffes were much taller than Shorty. This caused a problem during meal times, because Shorty's friends could eat the best leaves in the tall trees. Shorty was always hungry because she wasn't tall enough to reach the tall trees.
3. Ask the class how Shorty can solve her problem. Encourage the children to be creative. Explain to the class that you will only list the new ideas on the blackboard. Write each child's name next to his or her suggestion.
4. Give points for the most unusual solutions for Shorty's problem. Have the class vote on their favorite ideas.
5. SHORTY'S DIET: Draw a shorter tree on the blackboard with 7 leaves. If Shorty can reach 4 leaves, how many are left on the tree? (3 leaves) Challenge the children to write word problems to share with the class.

6. SHORTY TALES: Ask the children to write simple stories about how Shorty solved her problem. Encourage several children to share their solutions with the class. Write the children's ideas on the blackboard.

7. SHORTY'S NECK: If Shorty's neck was 4 feet (1.2 m) long and Shorty's body was 7 feet (2.1 m) tall. How tall was Shorty? (11 feet (3.4 m) tall) If the lowest branches of the tall trees are 15 feet (4.6 m) high, how much taller does Shorty have to be to eat from the tall trees? (4 feet (1.2 m))

8. SHORTY'S GANG: Ask the children if there are other animals that could be affected by being too short or too tall. Write the children's answers on the blackboard. Ask the children how these animals have adapted to their surroundings.

9. SHORTY'S TRIP: Draw the picture below on the blackboard. Tell the class the following story: Shorty decided to go for a walk. On the walk, Shorty found 4 short trees with 6 leaves on each tree. Shorty could only eat 12 leaves on this trip.

Ask the children to figure out how many leaves Shorty left on the trees. (12 leaves) If Shorty ate 3 leaves from each tree, how many leaves were left on each of the 4 trees. (3 leaves on each tree) Encourage several volunteers to come up to the blackboard and explain their answers to the class.

AROUND THE WORLD

A Primary- or Intermediate-Grade Activity
Stressing Brainstorming and Decision-Making

1. AROUND THE WORLD: Draw the above picture on the blackboard.

2. Explain the following situation to the class: Jasmine's family had been planning a trip to Africa for over 6 months. They planned on visiting Kenya, Zaire, and Tanzania. Jasmine's family flew into Nairobi, Kenya. Jasmine saw a sign in the airport that said "JAMBO." She asked her father what the sign meant. He said "Jambo means hello in the Swahili language." Later that week, Jasmine's family went on a photo safari to take pictures of the African wildlife. The safari truck's engine over heated near a Masai village. The leader of the safari tried to explain to the villagers that the truck needed water. Unfortunately, the Masai villagers didn't speak English.

3. How will the safari leader communicate with the Masai villagers if they don't speak the same language?

4. Ask the children if they were the safari leader, how they would explain to the Masai villagers that the safari truck needed water. Write the children's ideas on the blackboard. Encourage the children to share how they would communicate with someone who didn't speak English.

5. **TRIP AROUND THE WORLD:** Divide the class into groups of four or five children. Have each group plan a trip around the world visiting at least four or five countries. Decide how many days to spend in each country. It takes one day to travel from country to country. On the blackboard list each group's travel plans and the length of their stay in each country. Remind children they have only 21 days to travel, including the travel day between each country.

6. **AROUND THE WORLD TRAVEL:** Divide the class into groups of four or five children. Have each group call local travel agents to find the cost of a round trip flight to one of the countries on their around the world trip. Write the prices from several airlines on the blackboard. Have the children compare and determine which airline is offering the best price.

7. **AROUND THE WORLD AIRPLANE:** As a class, plan a trip around the world. Have the children choose five or six countries to visit. List the countries on the blackboard chart. Ask a local travel agent for help in planning this trip. Record the flight information, cost of the flights, and any changes in scheduling due to flight availability on the blackboard chart. Discuss all the details that must be taken care of when planning a trip.

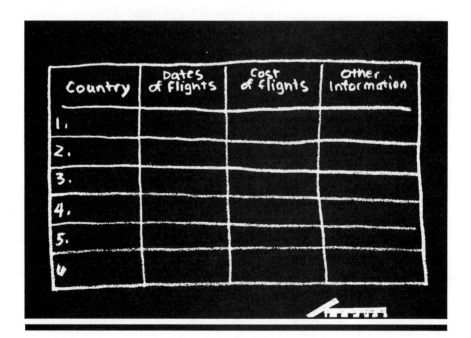

TREE FORT

**A Primary- or Intermediate-Grade
Activity Providing Brainstorming and Prioritizing Skills**

1. TREE FORT: Draw the above picture on the blackboard.
2. Explain the following situation to the class: Robert and Cheryl have found the perfect tree for building a tree fort. They have one problem, neither Robert nor Cheryl can reach the lowest tree branch.
3. Encourage the class to think of ways to reach the tree branch, bring the tools and materials up into the tree, and build the tree fort. Record the children's ideas on the blackboard. It is important to list all the suggestions, even though some of the ideas may seem silly.
4. Have the class vote on the best suggestions.
5. Encourage the children to combine ideas to solve the problem. Write the solutions on the blackboard.
6. TOWERING TREE FORT: Explain that the distance from the ground to the lowest branch is 6 feet (1.8 m). The children found a rope ladder in Robert's garage, but they are not sure how long it is. Cheryl measured 8 inches (20.3 cm) between each rung of the ladder. There

are 12 rungs on the rope ladder. How long is the rope ladder? (The ladder is 96 inches (2 m) long.) Is the ladder long enough to reach the bottom branch of the tree? (Yes, the tree branch is only 72 inches (1.6 m) high.)

7. TREE FORT CONSTRUCTION: Ask the children what materials and tools Robert and Cheryl will need to build their tree fort. Record the children's ideas on the blackboard. As a class, number the list of building materials and tools in order of importance. Discuss the results with the children.

8. TREE FORT DESIGN: Divide the class into teams of four or five children. Encourage each team to design a tree fort. Provide a section of blackboard for each group to use to draw on. Invite each team to describe the features that makes their tree fort unique. Have the class vote on the most creative tree fort.

COMMERCIAL CONVERSION

An Intermediate Activity Encouraging Creativity, Idea Modification Skills, and Adapting Concepts

1. COMMERCIAL CONVERSION: Draw the above picture on the blackboard.
2. Read the following commercial to your students: How would you like to make your dinner tonight without any effort? In less than five minutes, you can have a complete three-course dinner on the table with the new DINE-O food machine. DINE-O prepares, cooks, serves, and even cleans up after the meal. You'll love the convenience. Hurry and buy the amazing DINE-O food preparation machine, today!
3. Ask students why they think the DINE-O machine will sell well. List the important features on the blackboard.
4. Encourage the students to take turns at the blackboard drawing their own versions of the DINE-O machine.
5. Give the students 10 minutes to copy their ideas on paper. Display the drawings on the blackboard. Have the class vote on the best drawings based on imagination and original design.

6. Have the students take turns explaining how their DINE-O machines work.

7. COMMERCIAL CONVERSION DESIGNS: Divide the class into groups of four or five students. Encourage each group to design their own machine. Have the students give their machines names and explain what the machines can do. Write the name of each group's machine on the blackboard.

8. COMMERCIAL CONVERSION HOT SALES: Have each group write and perform a commercial advertising their new machine. Encourage the students to use the blackboard to advertise their products.

9. COMMERCIAL CONVERSION ANTIQUES: Encourage each student to choose an invention from history, such as the telephone, the phonograph, the airplane, and so on. Record each student's invention on the blackboard. Invite students to design commercials or printed advertisements selling their inventions from history. Encourage the students to perform their commercials or display their advertisements in the classroom.

10. COMMERCIAL CONVERSION PROFITS: Each DINE-O machine sells for $100.00 but only costs $25.00 to make. How much profit will you make if you sell 10 machines at a 25% discount? ($500 profit) Write the students' answers on the blackboard.

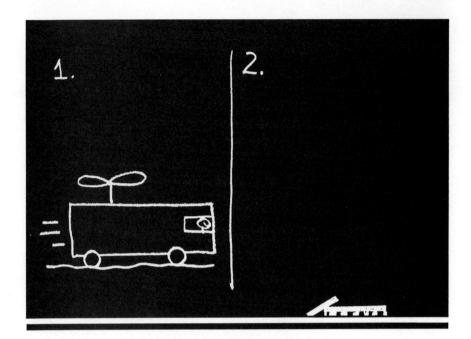

AIRMOBILE

An Intermediate Activity Encouraging Outcome Predicting, Graphic Recall, and Imagination

1. AIRMOBILE: Draw the above picture on the blackboard.
2. Give the students paper to draw what they think will happen next in picture #2.
3. After students have finished drawing and shared their ideas, draw your idea in the space for picture #2. For example, an airmobile flying through the clouds or above tall buildings.
4. THE FLYING AIRMOBILE: Ask the class what they would do with an airmobile if they could use it for 24 hours. Record the ideas on the blackboard. Encourage each student to write a story about his or her adventure. Have volunteers share their stories with the class.
5. TAKE A TRIP ON THE AIRMOBILE: Ask the students where they would fly if they had an airmobile. What kind of fuel does an airmobile use? Is an airmobile harmful to the environment? Write the students'

responses on the blackboard. Discuss the many possible uses of the airmobile.

6. THE AMAZING AIRMOBILE: Ask the students to think of the events that led up to the creation of the airmobile. For example, the inventor became tired of sitting in traffic. List the events on the blackboard. Ask the students to number the events in order of occurrence.

7. AIRMOBILE COMICS: Encourage the students to draw a 4-panel cartoon strip telling a story about the airmobile. Discuss the possible events that might take place in each cartoon panel. Record the students' ideas on the blackboard.

8. THE FLOATING AIRMOBILE: Ask students how they could make the airmobile float, too. Write the ideas on the blackboard. Encourage the students to think about where they would go if their airmobiles could fly and float. Refer the class to the book *Chitty Chitty Bang Bang* by Ian Flemming (Random House, 1964) for adventures of a car that could fly and float.

9. AIRMOBILE WEATHER: Explain the following situation to the class: The airmobile can travel at a speed of 75 mph (kmph) in clear weather and 50 mph (kmph) in bad weather.

How long will it take the airmobile to travel 400 miles (643.6 km) through 2 hours of stormy weather? (6 hours) Write the answers on the blackboard.

SHIPWRECKED

An Intermediate Activity Providing Opportunities for Prioritizing and Reinterpreting Problems and Solutions

1. SHIPWRECKED: Draw the above picture on the blackboard.

2. Explain the following situation: Your team has been shipwrecked on a desert island. The island is covered with rocks, trees, bushes, and fresh water. You must build a shelter that will protect your whole team before it gets dark.

3. Divide the class into four teams. Give each team newspaper and 12 inches (30.5 cm) of tape to build a shelter. Remind the students that they only have 10 minutes to work before you turn off the lights. If the teams are not finished building before the lights go out they will have to work in the dark.

4. Give the teams 10 minutes to build their shelters. After 10 minutes, turn off the lights and make the room as dark as possible.

5. After five minutes of "darkness" turn the lights on and compare each team's shelter. The best shelter is the one that will hold the entire team.

6. SHIPWRECKED SUPPLIES: Discuss what the students would do if they were shipwrecked on a desert island. Have the students name the supplies that would be most important to survival, such as matches, food, clothing, and so on. List the supplies on the blackboard in order of importance.

7. SHIPWRECKED SHELTER: Encourage the students to design shelters for the desert island. Ask the students what materials they would need to build the shelter. List the building supplies on the blackboard. What materials would you be able to recover from the shipwreck? Where would the other materials be found on the island?

8. SHIPWRECKED RESCUE: Divide the class into teams of four or five students. A ship has been spotted passing by the island. Ask the students how they would try to contact the ship. Have each team draw a picture on the blackboard showing how they would contact the passing ship.

9. SHIPWRECKED COCONUT: Coconuts are a good source of food and liquid. Most of the coconut trees are at least 10 feet (3.6 m) tall. Draw a picture of a coconut tree on the blackboard. Ask the students how they would reach the coconuts. Record the students' ideas on the blackboard. Encourage several students to draw pictures of how they would reach the coconuts.

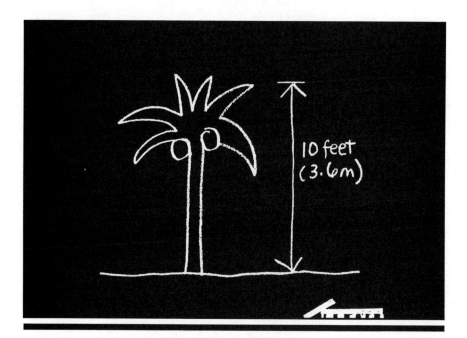

THE KID NEXT DOOR

An Intermediate Activity
Encouraging Original Ideas and Sequencing

1. THE KID NEXT DOOR: Draw the above picture on the blackboard.

2. Tell the students the following story: Maria wanted to play a game of checkers. Maria lives in the building next to her friend, Carlos. Maria can see Carlos through the window and motions to him to come and play checkers with her. Carlos arrives at Maria's door 24 minutes later.

3. Ask students why they think it took Carlos so long to arrive at Maria's house. Write the students' ideas on the blackboard. Record each student's name next to his or her idea.

4. Explain that you are giving points for the most unusual ideas. For example, Maria and Carlos live next door to each other in two 35-floor apartment buildings with broken elevators. This idea might earn as many as 5 points. The student with the most points wins.

5. **THE TEAM FRIEND NEXT DOOR:** Divide the class into four teams. Encourage each team to discuss and agree on several possible reasons for Carlos's delay. Have a spokesperson share his or her team's ideas. Record the responses on the blackboard. Have the class vote on the most creative ideas.

6. **THE LOFTY KID NEXT DOOR:** Explain that Carlos lives on the 30th floor of his apartment building. Each apartment is 15 feet (5.5 m) high and there is 3 feet (91.5 cm) between each floor. The lobby of the building is considered the first floor. How many feet is it from Carlos's apartment to the ground? (540 feet (16.5 m)) Record the students' answers on the blackboard.

7. **WHERE'S THE KID NEXT DOOR:** Explain the following situation to the class: Carlos and Maria both live on the 30th floors of their apartment buildings. The elevator is broken in Carlos's building. If there are 10 stairs per landing and 2 landings per floor, how many stairs are there between Carlos's apartment and the building lobby? (600 stairs) If it took Carlos 2 seconds a stair, how many minutes did it take Carlos to reach the lobby? (20 minutes) Write the answers on the blackboard. Encourage the students to write challenging word problems for their classmates.

8. **TALKING WITH THE KID NEXT DOOR:** Discuss other ways Maria and Carlos could have communicated. Write the students' responses on the blackboard. Encourage several students to draw their ideas on the blackboard.

9. **A QUIET KID NEXT DOOR:** Divide the class into pairs. Have each pair try to communicate an idea without talking, such as hunger, fatigue, fear, and so on. Discuss how the students were able to communicate. Record the responses on the blackboard. Ask several pairs of students to share with the class how they were able to communicate with each other.

THE QUEEN'S GUARD

An Intermediate Activity Providing
Experience in Adapting and Deduction

1. THE QUEEN'S GUARD: Draw the above picture on the blackboard.
2. Tell the class the following story: The queen had the most loyal guard in the kingdom. She was constantly frustrated with his behavior because he would obey every command but one. No matter how many times she asked the guard to bring her a certain object, he would always refuse to obey.

   Ask the students to list objects that the guard might refuse to bring to the queen. Write the ideas on the blackboard. Record each student's name next to his or her idea. Encourage the students to be creative.
3. Explain that points will be awarded to the most creative and unusual ideas. The student with the most points wins. For example, the guard is actually a dog that refuses to bring the queen her slippers.
4. PROTECTION FROM THE QUEEN'S GUARD: Divide the class into groups of four or five students. Explain the following situation: The queen owns a valuable jewelled crown. She wants the crown to be guarded 24 hours a day. What's the best way to protect the crown?

Give the teams 30 minutes to design a burglar alarm. Invite one member from each team to draw the design on the blackboard and explain how the burglar alarm works. Have the class vote on the best design.

5. HIDE THE QUEEN'S GUARD: Divide the class into teams of four or five students. Send one team out in the hall. Have the rest of the students choose an object to hide in the classroom. Bring the team back in the room. Have the other students give clues describing the object that is hidden. Give the teams 5 minutes to figure out the object that is missing. Write each team's guesses on the blackboard. Give each team a turn in the hall. The team with the fewest wrong guesses wins.

6. THE QUEEN'S GUARD DOGS: Draw the picture of the two dogs on the blackboard. Explain the following situation to the class: The queen has two dogs, Willie and Billie. She has had the dogs since they were puppies. Willie is 84 dog years old and Billie is 56 dog years old. Point out that one human year is equal to seven dog years. If the queen is 35 years old, how old was she when she got each puppy? (The queen was 23 years old when she got Willie and 27 years old when she got Billie.) How old are the dogs in human years? (Willie is 12 years old and Billie is 8 years old.) Write the students' answers on the blackboard. Encourage students to change the word problems to make them more challenging. Invite students to share their new word problems with the class.

ACROSS THE TRACKS

An Intermediate Activity Providing Opportunities for Idea Inventing, Sequencing, and Reinterpretation

1. ACROSS THE TRACKS: Draw the above picture on the blackboard.
2. Tell the following story: Every morning, Mr. Adams walks to work. He crosses tracks on the way but he never worries about oncoming trains.
3. Ask students why Mr. Adams never worries about oncoming trains. Write the students' ideas on the blackboard. Record each student's name next to his or her idea.
4. Give points for the most unusual ideas. For example, the tracks might be toy train tracks on the living room floor or dog tracks on a muddy road.
5. MYSTERIES ACROSS THE TRACKS: Discuss the different kinds of tracks Mr. Adams could be crossing each day, such as the toy train tracks, dog tracks, or car tracks. List the ideas on the blackboard. Ask students to write stories about Mr. Adam's walk to work.

6. ADVENTURES ACROSS THE TRACKS: Mr. Adams leaves for work at 7:00 am each morning. He walks to work and arrives at 8:00 am. What does Mr. Adams do on his walk to work? List the events in order on the blackboard. Discuss why Mr. Adams walks to work everyday.

7. TEAM ACROSS THE TRACKS: Divide the class into four teams. Have each team discuss and agree upon possible solutions to why Mr. Adams doesn't worry about oncoming trains on his way to work. Have one person from each team share their team's best answer. Write the responses on the blackboard.

8. WALK ACROSS THE TRACKS: Write the following problem on the blackboard: It takes Mr. Adams 15 minutes to walk 1 mile (1.6 km). If it takes Mr. Adams 1 hour to get to work, how far does Mr. Adams walk to work? (Mr. Adams walks 4 miles (6.4 km) to work.) How long would it take Mr. Adams to walk 5 miles (8.0 km) if he took a 10-minute break every 2 miles? (It would take Mr. Adams 1 hour 35 minutes to walk 5 miles (8.0 km).)

OUT THE DOOR

An Intermediate Activity
Encouraging Original Idea Development

1. OUT THE DOOR: Draw the above picture on the blackboard.
2. Tell the class the following story: Every day Mrs. Clark leaves for work out the front door of her house. At the end of the day, she always enters the house through the back door.
3. Ask the students why Mrs. Clark leaves her house using one door and returns home using another door. Record the students' ideas on the blackboard.
4. Give points for the most original ideas. Write each student's name next to his or her answer. For example, Mrs. Clark is handicapped and in a wheelchair. Mrs. Clark uses the ramp from her front door to leave the house and the area around the back door is on level ground with easy wheelchair access.
5. OUT THE DOOR JOB: Encourage the students to discover Mrs. Clark's occupation. Give the students the following information: Mrs. Clark wears a uniform to work. She sometimes wears a hat at work. What does Mrs. Clark do? Record the students' ideas on the blackboard. Encourage the students to not limit their ideas to stereotypical female roles.

6. SUDDENLY OUT THE DOOR: Write the following facts on the blackboard—Mr. Clark leaves for work at the same time each day, he enters the building through the front door but is never seen leaving at the end of the day. Mr. Clark arrives home at the same time each night. Encourage students to write stories about Mr. Clark's daily adventures. Have the class agree on a particular occupation for Mr. Clark before writing their stories.

7. IN AND OUT THE DOOR: Discuss the sequence of events in Mr. Clark's day. List the events in order on the blackboard.

8. MILES OUT THE DOOR: Explain that Mr. Clark works 1 mile (1.6 km) away from home. At lunchtime, Mr. Clark walks 50 yards (45.7 m) to a park to eat his lunch and 50 yards (45.7 m) back. How many feet (m) does Mr. Clark travel to work, to lunch, and back home again each day? (Mr. Clark walks 10,860 feet (4.1 km) each day.) Remind the students that there are 5280 feet (1.6 km) in 1 mile (1.6 km). How many feet would Mr. Clark travel after 5 days of work? (Mr. Clark walks 54,300 feet (20.6 km) in 5 days.) Write the students' answers on the blackboard.

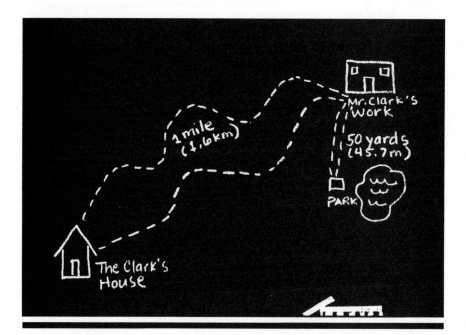

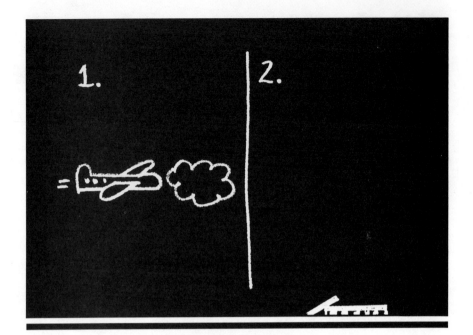

CLOUD NINE

An Intermediate Activity for Developing Original Thinking and Outcome Predicting.

1. CLOUD NINE: Draw the above picture on the blackboard.
2. Divide the class into four or five teams. There should be no more than six members on a team.
3. Give students sheets of paper and 5 minutes to draw what they think happened next in the scene on the blackboard.
4. Encourage each team to discuss and agree on the most original idea for what happened in picture #2.
5. Have each team send one member to draw their best idea on the blackboard. Encourage the teams to explain what is happening in their pictures. Have the class vote on the most original idea.
6. STORMY CLOUD NINE: Divide the class into groups of five or six students. Have each team research a cloud type—cumulus, cirrus, nimbus, and so on. Encourage each team to share information and draw examples of their clouds on the blackboard. Ask the students how the different clouds would affect airplanes.

7. CLOUD NINE'S FAREWELL: The airplane is traveling at 500 miles (840.5 kmph) per hour. If the plane still has 1200 miles (1930.8 km) to travel, how many hours and minutes will it take to reach its destination? (2 hours, 24 minutes) Record the students' answers on the blackboard.

8. CLOUD NINE TRIP: Divide the class into teams of four or five students. Provide each team with a world map. Ask the students to consider the following question: If you could take an airplane trip anywhere in the world where would you go? Explain that the airplane has enough fuel to fly 8 hours at 500 miles per hour (840.5 kmph). Encourage each team to plan a trip based on the amount of fuel. Use the scale on the map to help figure out the distance the airplane could fly. Write each team's final destination on the blackboard.

9. THE EVENTS OF CLOUD NINE: Ask the class why pilots file flight plans each time before their planes takes off. Discuss the events that might have taken place during Flight 809. Have the class agree on a departure location and a destination for the flight. Write the flight information on the blackboard. Encourage the students to write stories about the adventures of Flight 809. Explain that in each story all passengers must arrive safely at their destination. Ask volunteers to share their stories with the class.

THE CHASE

An Intermediate Activity to
Encourage Team Brainstorming and Elaboration

1. THE CHASE: Draw the above picture on the blackboard.
2. Give students sheets of paper and 5 minutes to draw what they think happened next in the scene on the blackboard.
3. Divide the class into four or five teams. Encourage each team to discuss and agree on the most original idea for the next event in picture #2.
4. Have each team send one member to draw their best idea on the blackboard. Encourage the teams to explain what is happening in their pictures. Have the class vote on the most original idea.
5. WHAT'S CHASING ME?: Encourage students to imagine what is chasing the person in the picture. Write the students' ideas on the blackboard. For example, insects, small spaceships, and so on. Have students write stories about the chase.
6. AVOIDING THE CHASE: Encourage each student to think about what he or she would do if he or she were being chased. Ask the students what's chasing them. How would the students get away

from the swarm and why are they are being chased? Write the answers on the blackboard. Discuss the variety of responses.

7. HOW THE CHASE HAPPENED: Discuss the events before, during, and after the chase. List the ideas on the blackboard. Have the students number the events in the order they occurred.

8. THE MYSTERY CHASE: Divide the class into teams of four or five students. Ask the teams to think about the following situation: You are being chased in a field by a bunch of children in uniforms.

   Ask the students why they are being chased. Encourage the teams to discuss possible answers to the question. For example, the student is carrying a football and is being chased by the opposing team. Have each team share one of their ideas with the class. Write the responses on the blackboard.

9. THE CHASE COMICS: Draw a five-panel cartoon strip on the blackboard. Encourage the students to design a cartoon about a chase. As a class, discuss what should be included in each cartoon panel. Record the students' ideas on the blackboard. Ask volunteers to draw the events in each panel.

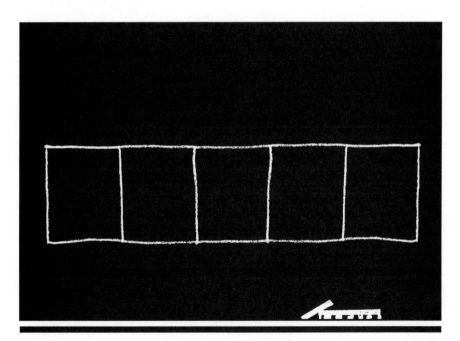

MUSICAL SEARCH

An Intermediate Activity for
Applying Evaluation Skills to Music

1. MUSICAL RESEARCH: Provide a tape recorder and several tapes featuring common instruments, such as trumpets, trombones, saxophones, piano, guitars, violins, flutes, drums, and so on.
2. Write the names of several instruments on the blackboard. The list should include more instruments than are played on the tape.
3. Divide the class into four or five teams. Play a musical tape that includes some of the instruments listed on the blackboard.
4. Have the teams listen to the music for 5 minutes and agree on the instruments they hear. The whole team must agree on a single list of instruments. Have each team read their list. Record the instruments on the blackboard.
5. The team that names the most instruments correctly on the tape, wins.
6. MUSICAL SEARCH AROUND THE WORLD: Divide the class into teams of four or five students. Play tapes of music from a variety of countries around the world. Challenge the teams to identify which

country each piece of music is from. Write the students' answers on the blackboard. The team that matches the most countries and music is the winner. Provide a world map for reference.

7. MUSICAL SEARCH SOUND EFFECTS: Ask the students how they can make musical instruments from combs, wax paper, tin cans, bottles, and so on. Write the list of materials on the blackboard. Encourage the students to create their own musical instruments using a variety of objects. Challenge the students to play "Twinkle, Twinkle" on their musical instruments.

8. MUSICAL SEARCH NOTES: Bring in eight glass soda-pop bottles, a gallon of water, and a 2-cup measuring cup. Draw eight empty bottles on the blackboard. Ask the students how to play a musical scale using only water and pop bottles. Write the students' ideas on the blackboard. Place the eight bottles in the front of the room. Pour 1/4 cup of water in the first bottle and 2 cups of water in the last bottle. Have a volunteer tap each bottle to hear the sound. What's the difference between the two sounds? Write the students' responses on the blackboard. Explain that high and low notes are made by pouring different amounts of water in the bottles. Challenge the students to make a musical scale. Encourage the students to experiment with the amounts of water in each bottle. As a class, arrange the bottles in order from the lowest to the highest notes.

9. MUSICAL SEARCH COMBOS: Discuss the types of music the students like to listen to. Have the students list the instruments used in their favorite bands. Write the answers on the blackboard. Ask the students to list the instruments they would want to include in their own bands. Write the students' responses on the blackboard.

10. MUSICAL SEARCH SHOW: Play different types of music for the students, such as jazz, blues, country western, classical, rock and roll, and so on. Challenge the students to identify each type of music. Write the students' answers on the blackboard chart.

	Music 1	Music 2	Music 3	Music 4	Music 5
jazz					
country					
blues					
classical					
rock & roll					

OBSTACLE COURSE

An Intermediate Activity for Individual and Team Problem-Solving and Creative Design of Physical Movement

1. OBSTACLE COURSE: Draw the above picture on the blackboard. Adapt the illustration to include equipment that is found on your school playground.
2. Give the students pieces of paper and ask them to copy the blackboard drawing.
3. Explain that you have created an obstacle course on the playground that is similar to the drawing on the blackboard. Point out that the dotted lines on the drawing represent wooden boards connecting the equipment. The object is to move a water balloon from one end of the obstacle course to the other end without the students touching the balloon with their hands. Explain that the students will be using different objects to help move the balloon, such as coat hangers, string, spoons, paper, and so on.
4. Have one student use a ruler, the next a coat hanger, the next a piece of string, and so on. Before trying the obstacle course, encourage the students to predict which tool will be the easiest to use with the balloon. Record the students' responses on the blackboard.

5. SPEEDY OBSTACLE COURSE: Keep track of the time it takes each player to finish the obstacle course. Record the students' times and tools on the blackboard. The student who moves the balloon across the obstacle course in the least amount of time is the winner. Ask the students which is the best tool to use on the obstacle course. Encourage the students to explain their answers.

6. OBSTACLE COURSE ROLL: Divide the class into teams of three or four children. Encourage each team to design a challenging obstacle course for a water balloon. Give the teams 45 minutes to work. Have each team draw their obstacle course on the blackboard. Ask the students to vote for the most creative design.

7. OBSTACLE COURSE TEST: Divide the class into three teams. Give each team a ruler and a different object to move through the classroom obstacle course, such as a tennis ball, a hard-boiled egg, and a marble. Encourage the students to predict which item will be the easiest and the most difficult to maneuver through the course. Record the students' predictions on the blackboard chart. Have the teams test their predictions. Compare the actual results with the class predictions. The team that predicted correctly wins.

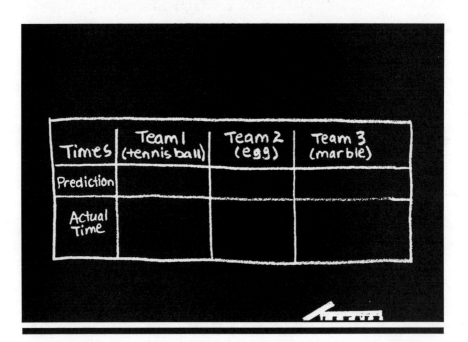

8. TEAM OBSTACLE COURSE: Divide the class into four or five teams. Each team member must use a different tool— a spoon, a piece of string, a coat hanger, and a piece of paper— to move the balloon through one obstacle on the course. The team that finishes the obstacle course in the shortest amount of time is the winner. Record the times on the blackboard.

9. INSIDE OBSTACLE COURSE: Divide the class into teams of four or five students. Have each team design and build an obstacle for a classroom obstacle course. Challenge each team member to use a ruler to move a ping-pong ball through one obstacle on the course. The team that finishes the entire obstacle course in the shortest amount of time is the winner. Record each team's time on the blackboard chart.

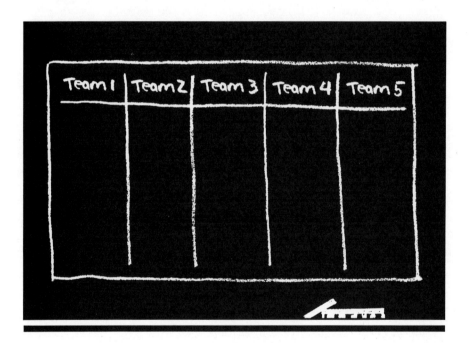